For my children, Alia, Zarena, Zaki and Jaan,
and for all the doctors and nurses and healthcare workers
who look after every bit of our brilliant bodies. **R.F.**

To all children with curious minds. The more we learn,
the more we will want to learn. **V.W.**

First published 2024 by Walker Books Ltd
87 Vauxhall Walk, London SE11 5HJ

2 4 6 8 10 9 7 5 3 1

This book has been typeset in Mali

Printed in China

British Library Cataloguing in Publication Data: a catalogue record for this book is available from the British Library

ISBN 978-1-5295-0452-1

www.walker.co.uk

Dr Roopa's Body Books

THE SUPER SKELETON

DR ROOPA FAROOKI

illustrated by

VIOLA WANG

WALKER BOOKS
AND SUBSIDIARIES
LONDON • BOSTON • SYDNEY • AUCKLAND

What do we have in common with fish and snakes and birds? We all have **skeletons**: the **bones** that give our bodies their shape. If you didn't have a skeleton, you'd just be a squashy blob ... like a jellyfish!

Your skeleton does all kinds of different jobs. Whenever you move your body – whether that's walking, dancing, skipping or swimming along – you are using your skeleton.

Your skeleton also protects the really important bits of your body, like your brain, lungs, heart and spinal cord.

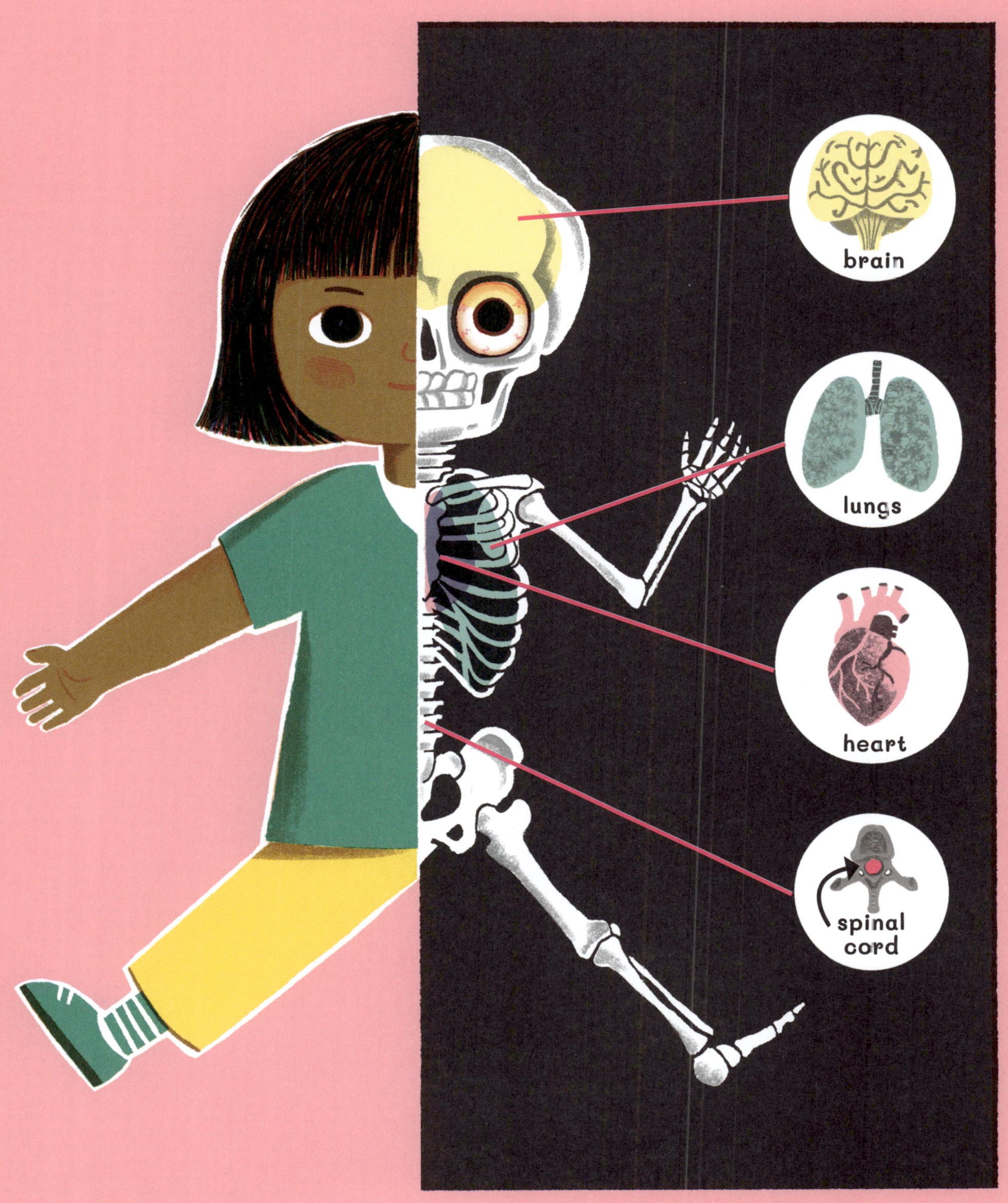

To grow strong, healthy bones, your body needs something called **calcium**. The outer layer of your bones is made from calcium – and this is the hardest part of your body!

You get calcium from foods like milk, cheese and yoghurt. And you need Vitamin D to absorb it, which you mainly get from sunshine.

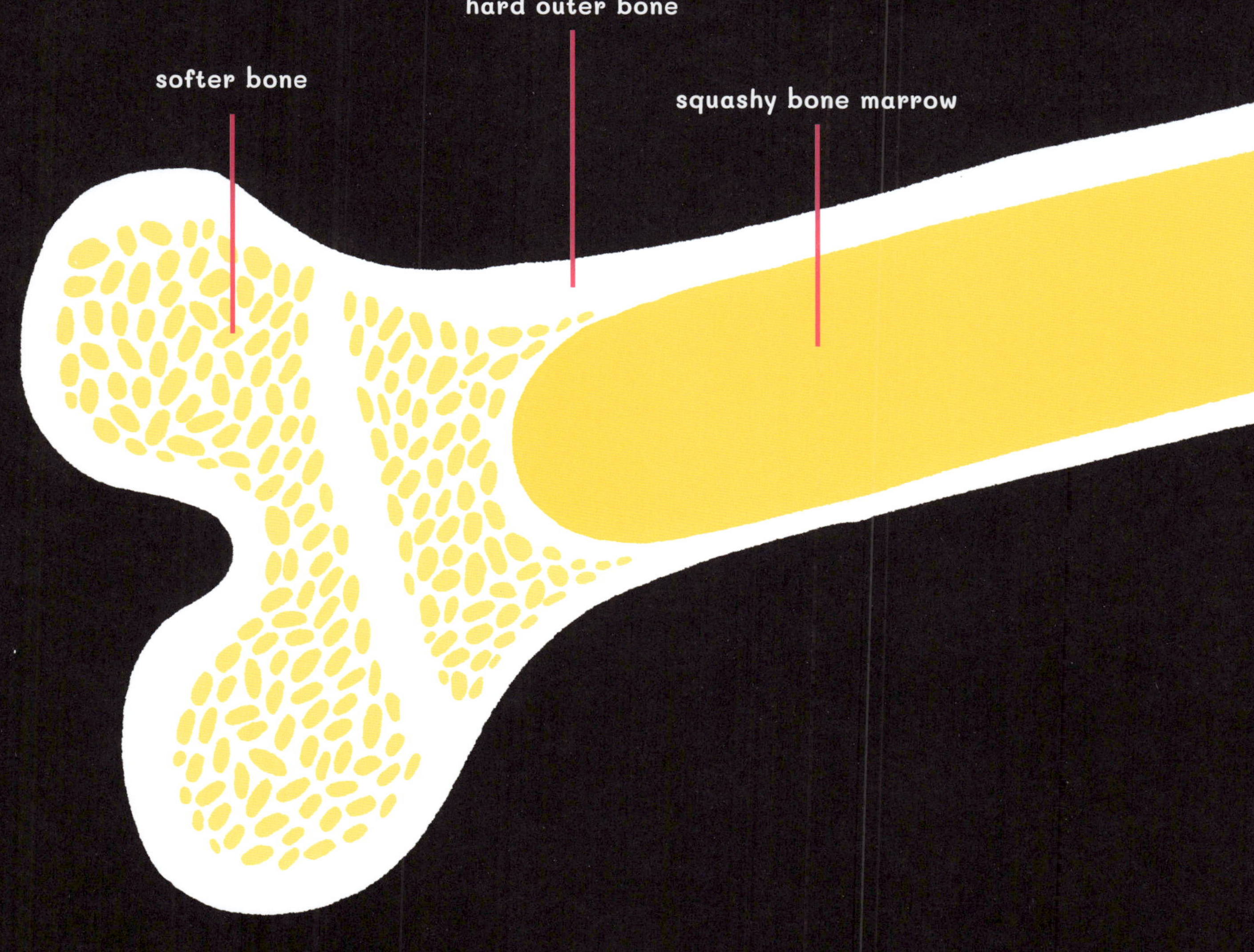

Beneath the hard outer surface is a softer layer and, inside that, bones have a wet and squashy centre made of stuff called "bone marrow".

Bone marrow is essential for making blood, which takes the oxygen and energy you need to every cell in your body.

Your skeleton can be divided into two sections:

1) The **central skeleton** contains the bones that run down the middle of your body, like your SKULL, SPINE and RIBCAGE.

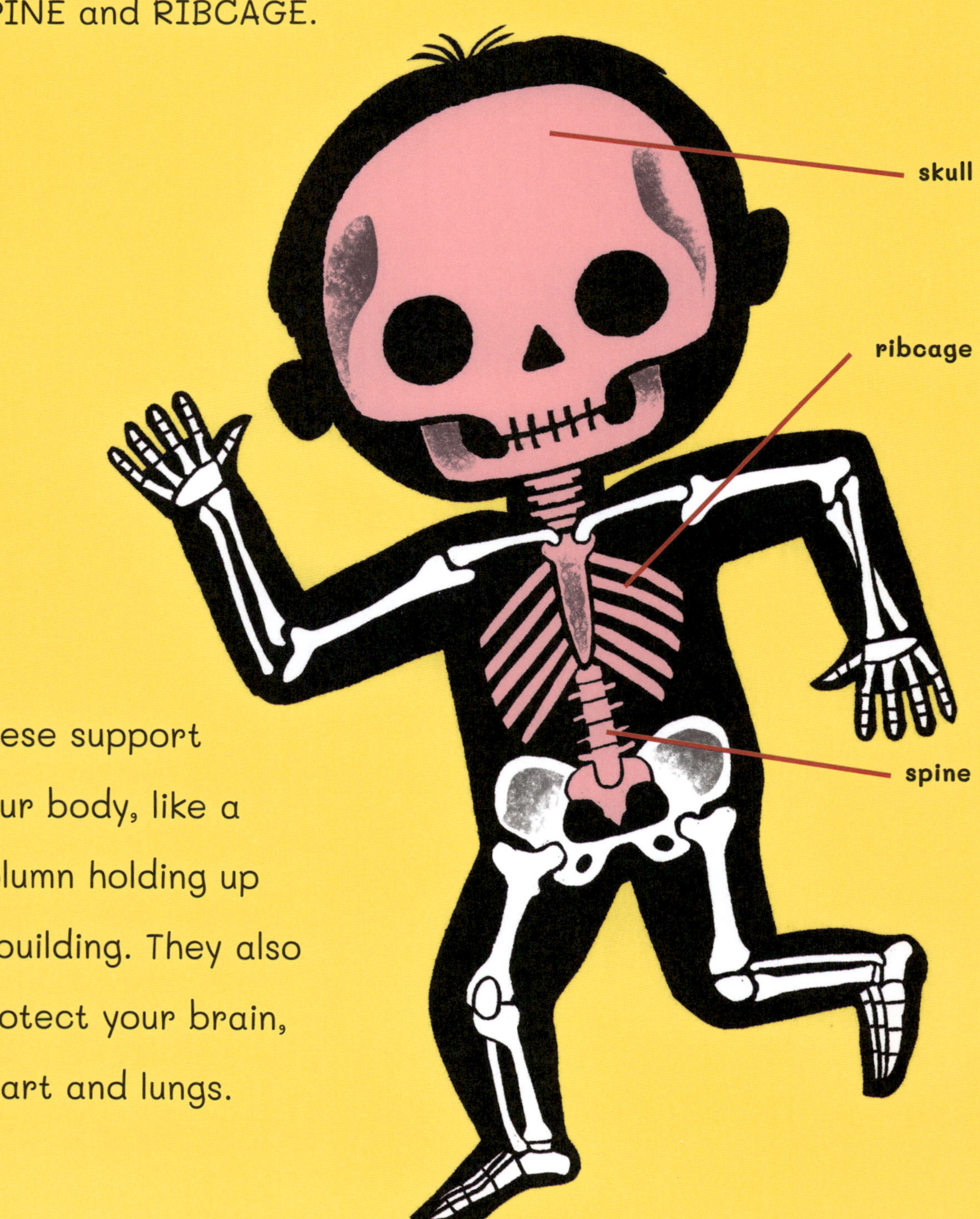

These support your body, like a column holding up a building. They also protect your brain, heart and lungs.

2) Other bones hang off the central skeleton, like the ones in our arms and legs. These are called **appendages**, which means "extra bits"!

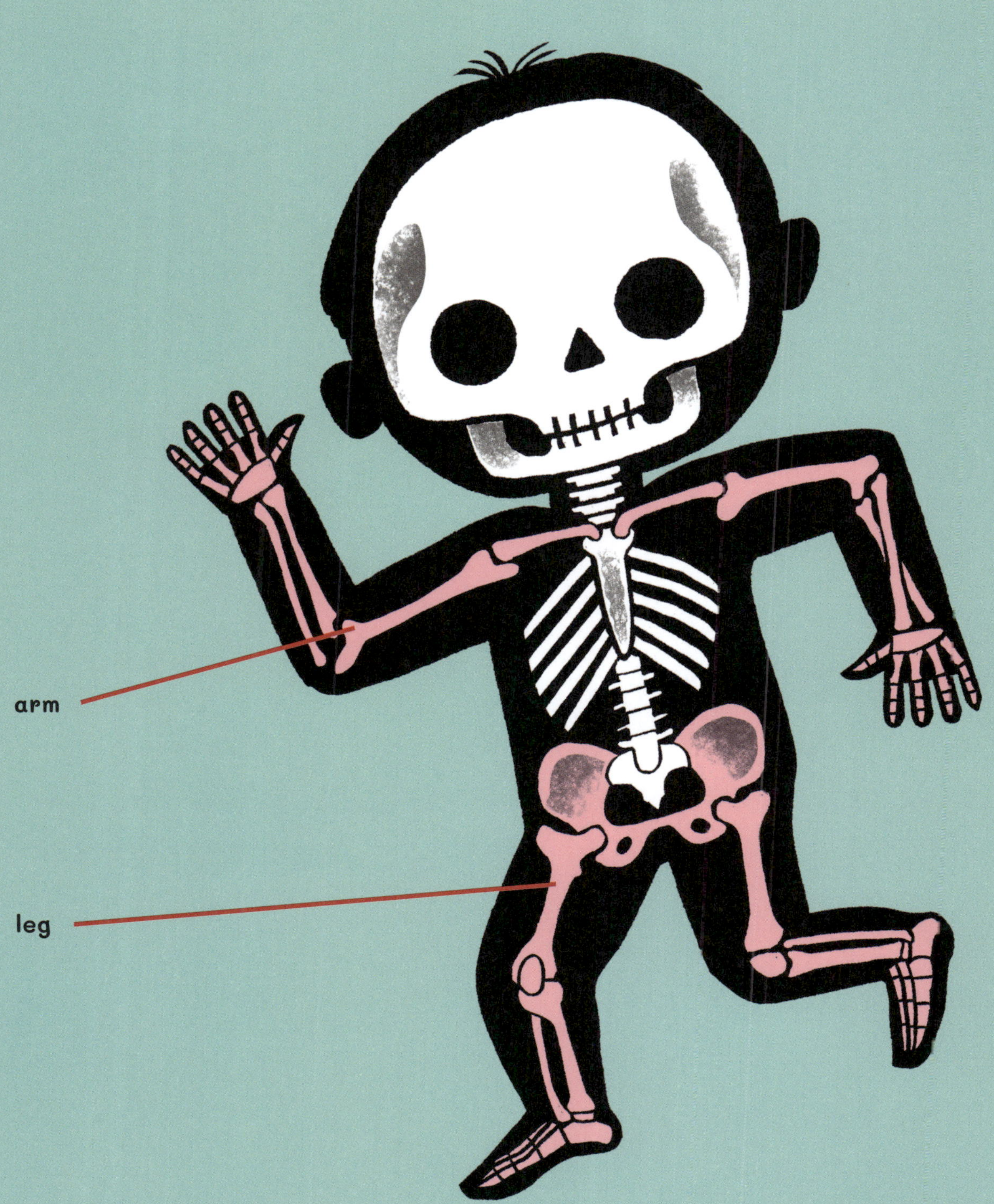

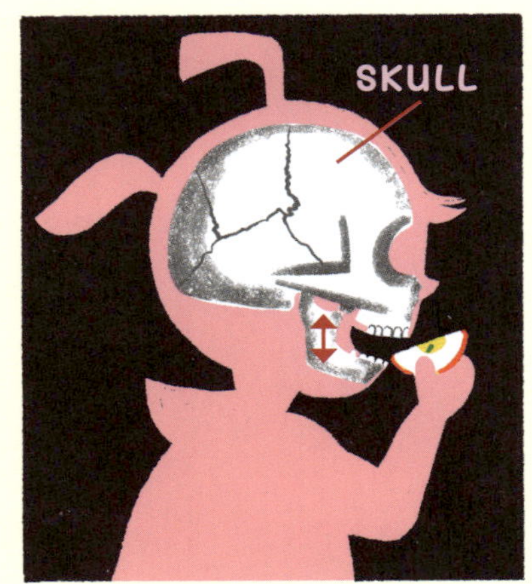

The most important bones are probably the ones in the **SKULL**. The skull protects the brain, which is like the control room for the rest of the body.

The skull is connected to the jaw. This is made up of a pair of bones. The upper part is fixed but the lower part can move up and down, which allows us to eat and talk!

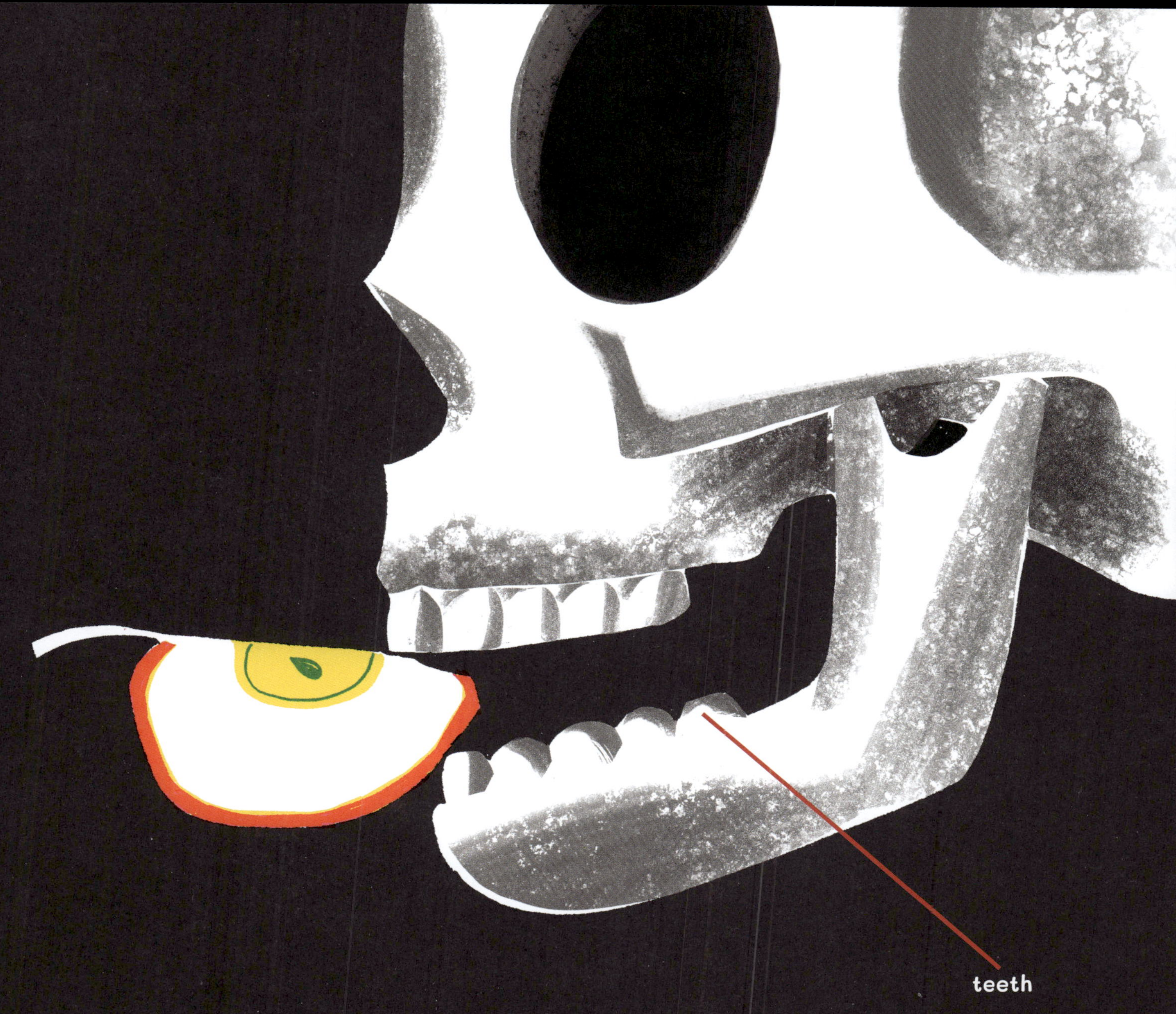

Held in the jawbones are the teeth. They look like bones — they're white and also made of hard calcium — but they actually aren't... They can't heal themselves when they are broken, whereas bones can. (More on that soon!)

The spine has a very important job too. It protects the spinal cord: the bundle of nerves in the back which sends signals to and from the brain.

The spine gets shorter during the day, because the softer discs between the bones squash when someone stands and sits... So you are likely to be a bit smaller by bedtime!

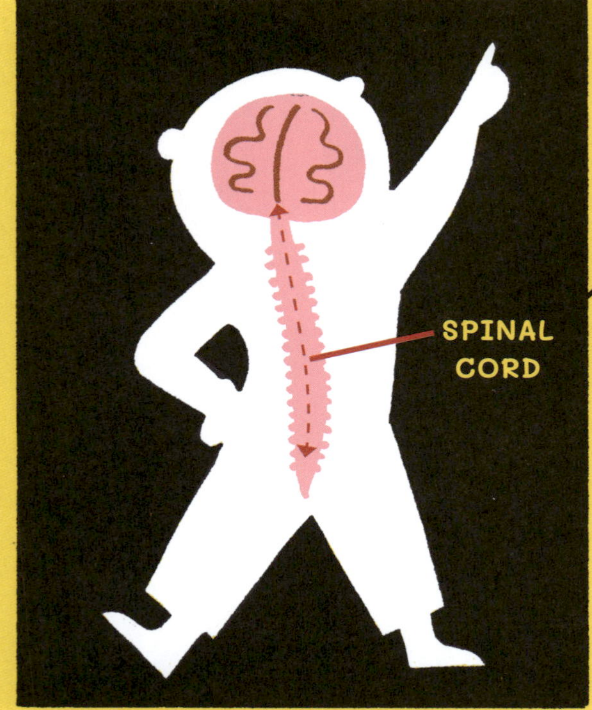

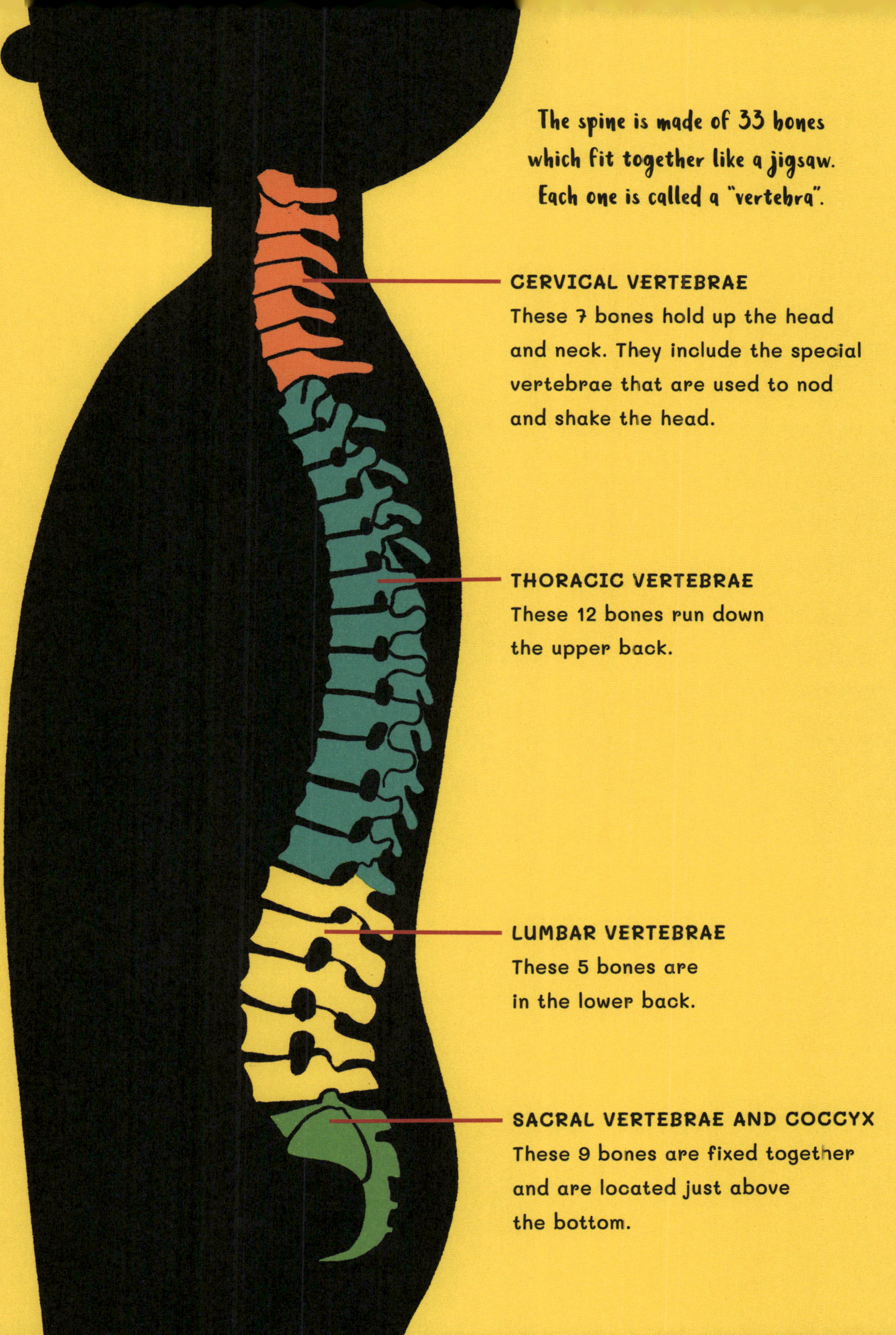
The spine is made of 33 bones which fit together like a jigsaw. Each one is called a "vertebra".
CERVICAL VERTEBRAE
These 7 bones hold up the head and neck. They include the special vertebrae that are used to nod and shake the head.
THORACIC VERTEBRAE
These 12 bones run down the upper back.
LUMBAR VERTEBRAE
These 5 bones are in the lower back.
SACRAL VERTEBRAE AND COCCYX
These 9 bones are fixed together and are located just above the bottom.

Your bones are connected with JOINTS. Some joints don't move at all, like the bones in the middle of your ribcage (called the "sternum"), which are fixed together.

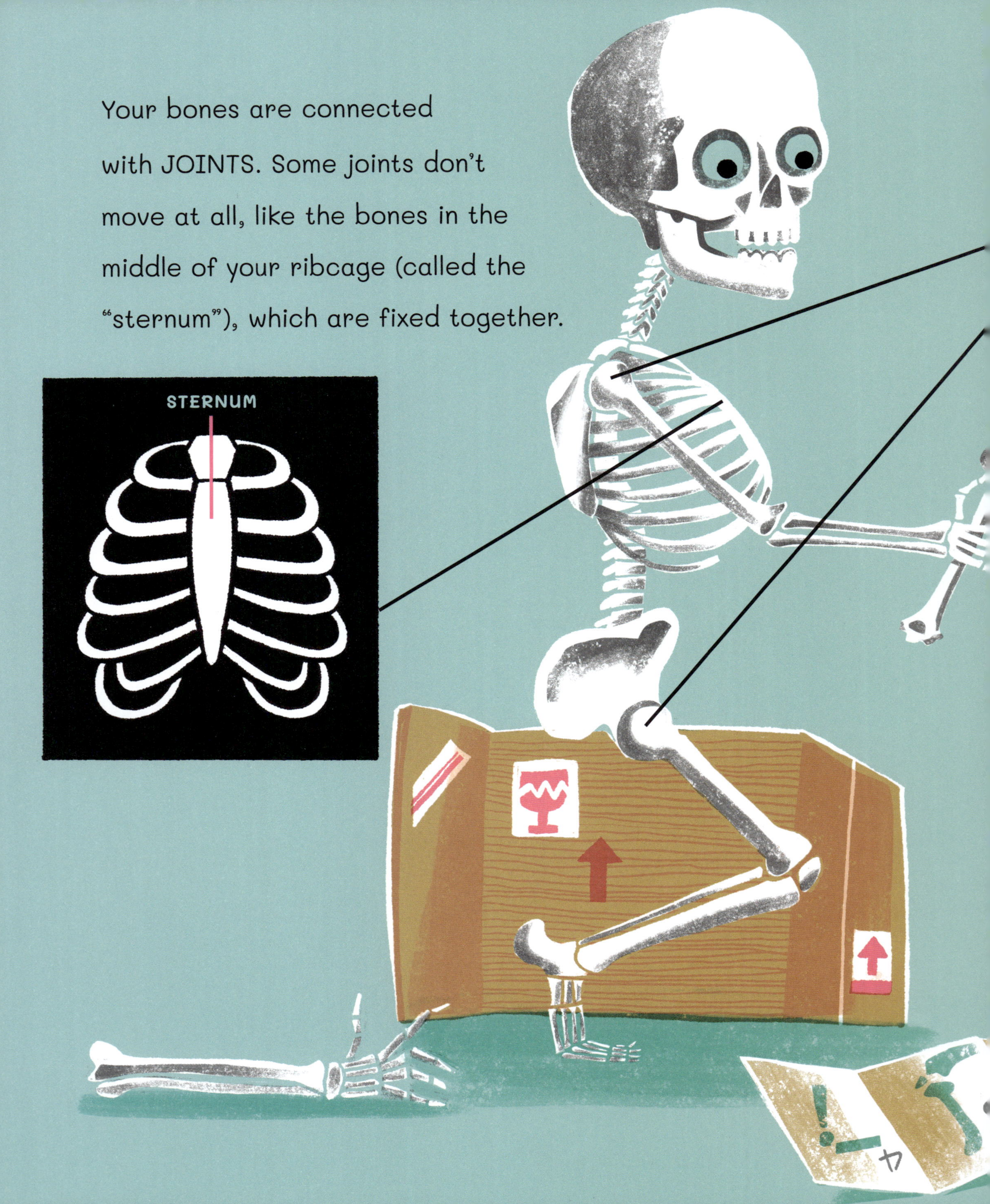

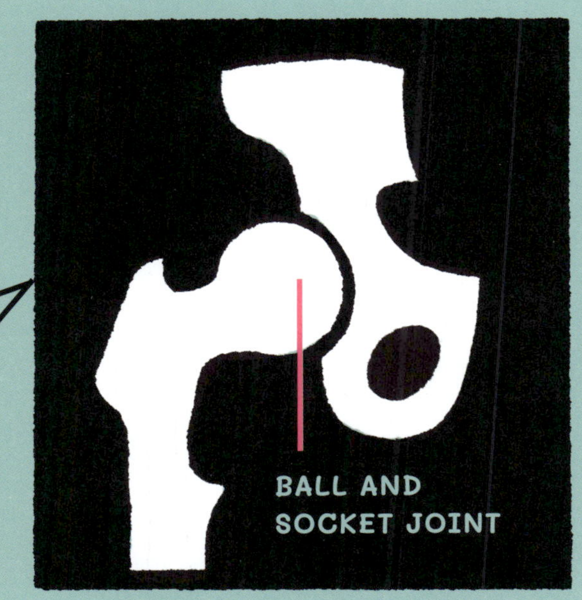

Some let the bones move a lot. One example is the ball and socket joint, which is in the hips and shoulders. It swivels so that the arms can swing round like windmills!

HINGE JOINT

There are also hinge joints, in the knees, elbows and jaw, which work like the hinges in doors.

Muscles connect to bones. When someone wants to move a body part, a signal comes from the brain to the nerves, and then to the muscles.

When the signal arrives, the muscles will bunch up, getting shorter and pulling on the bone they're attached to ... or relax and get longer, letting the bone move back to where it was.

Sometimes, bones can move without you meaning them to! If a doctor bangs someone's knee, their leg can jerk up — or if someone touches something hot, their hand will jerk away. This is a "reflex": a quick message to the spine and back to the muscle which doesn't even reach the brain.

Babies have more bones than adults! We're born with about 270 bones and have about 206 by the time we're grown-ups. This is because some bones, like the longer ones in our arms and legs, are separate when we're babies, and then join together as we get older.

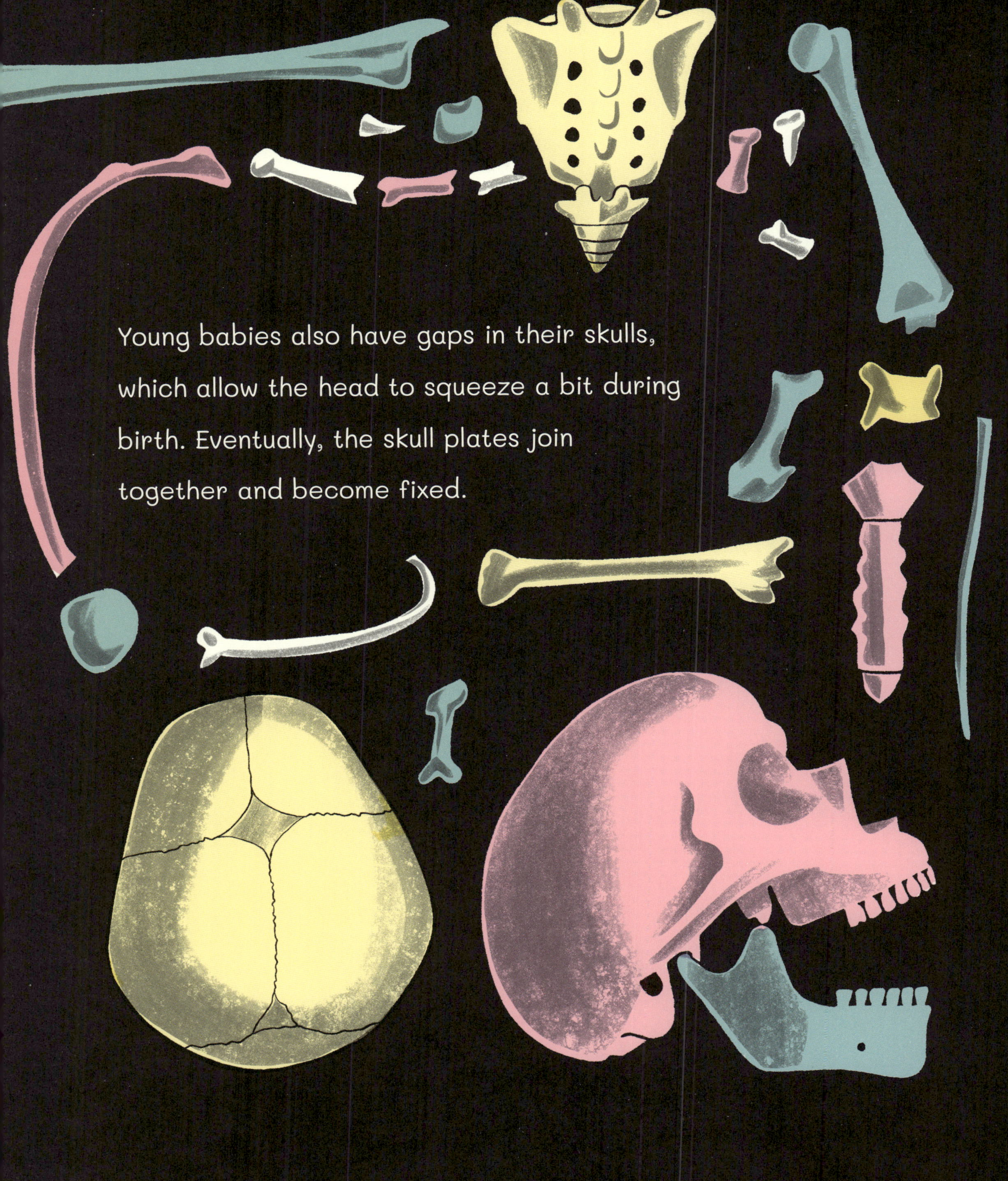

Young babies also have gaps in their skulls, which allow the head to squeeze a bit during birth. Eventually, the skull plates join together and become fixed.

What happens if you break a bone? Let's say someone breaks an arm. Ouch! It hurts to move and swells up.

The first step is to go to the hospital, to a place called A&E (which stands for "Accident and Emergency"). A doctor will examine their arm and give medicine if it hurts. Then ...

they will arrange for an X-ray. An X-ray creates a special photo that looks right inside the body.

The bone shows up as bright white, so the doctors can tell if there is a break.

The broken bit might need to be kept in place with a plaster cast.

This means the bone can grow back together in the right place.

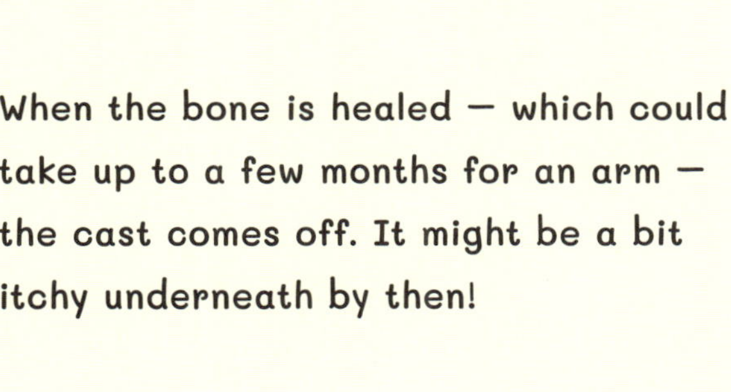

When the bone is healed — which could take up to a few months for an arm — the cast comes off. It might be a bit itchy underneath by then!

What's the most special thing
about our super skeletons?
They let us move our bodies
in lots of wonderful ways,

whether that's talking, waving, dancing ... however we express ourselves. It's the skeleton that helps us to get up if we fall – and give friends a helping hand, too!

AUTHOR'S NOTE

My name is Doctor Roopa Farooki, and I've helped look after every bit of people's brilliant bodies, from their brains to their big toes!

There are all kinds of different ways to look after your bones, muscles and joints, to keep yourself active and feeling great. Here are my very top tips for looking after our super skeletons:

- **Eat a good mix of food, including products with plenty of calcium (dairy foods like milk, cheese and yoghurt, or dairy alternatives), to help you grow strong bones.**

☆ Try to spend at least 20 minutes a day outside, to get a good dose of Vitamin D. This helps you absorb the calcium that builds your bones.

☆ Drink lots of water! You need to be well hydrated to exercise and keep your bones working properly.

☆ Move your body when you can. Some people do groovy dance moves, some wiggle their toes and wave their hands... So work out which way works best for you.

☆ Protect your skeleton when you're being active, whether that's on a bike or playing sports – wear a helmet or pads, where recommended. Have fun safely!